BIONICS

WRITTEN BY

MAXINE ROSALER

BLACKBIRCH®
PRESS

THOMSON
★
GALE

San Diego • Detroit • New York • San Francisco • Cleveland • New Haven, Conn. • Waterville, Maine • London • Munich

LIBRARY OF CONGRESS CATALOGING-IN-PUBLICATION DATA

Rosaler, Maxine.
    Bionics / by Maxine Rosaler.
        p. cm. — (Science on the edge)
Summary: Discusses the history of replacement body parts, current accomplishments in the field, and visions of future technology.
Includes bibliographical references and index.
    ISBN 1-56711-784-8 (hardback : alk. paper)
    1. Artificial limbs—Juvenile literature. 2. Bionics—Juvenile literature. [1. Bionics. 2. Prosthesis. 3. Artificial organs.] I. Title. II. Series: Science on the edge series.

RD756 .R765 2003
617.9'5—dc21   2002015970

TABLE OF CONTENTS

Machines have some advantages over human beings. Their parts can be replaced, sometimes in ways that are improvements on the original. When a car's battery dies, a new, stronger one can be installed. A regular camera lens can be replaced with a zoom lens if a photographer wants to take close-up pictures of something far away. It is difficult to imagine humans with this ability to exchange their old, used up, or missing parts with new ones that can do the same job.

Yet today, to an increasing degree, human beings can do just that. Modern medical technology has made it possible to replace many parts of the human body with artificial devices that work as well, or even better, than the original. This technology, which blurs the line between humans and machines, is called bionics.

To some people, the word "bionics" conjures up a futuristic image of superhumans with radar vision and computer-chip memory. To others, it is simply a description of medical devices that are part of everyday life. Bionic devices range from artificial arms and legs to hearing aids and replacements for the human heart.

Bionic devices have improved a great deal in the last few decades. They will continue to be refined, thanks to advances in miniaturization, development of new materials and computers, and increased understanding of the human body. Within decades, it is possible that bionics will enable blind people to see with the help of implanted cameras. Amputees may control artificial limbs directly with their nerves. Eventually, bionics may give people abilities never before thought possible, such as the ability to control distant machinery with the power of thought.

Bionics is the science of using machines to help the human body.

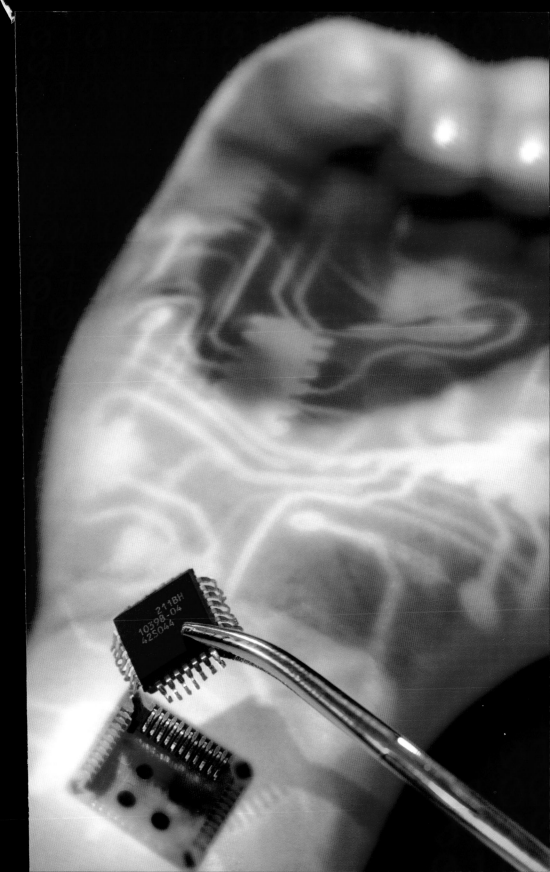

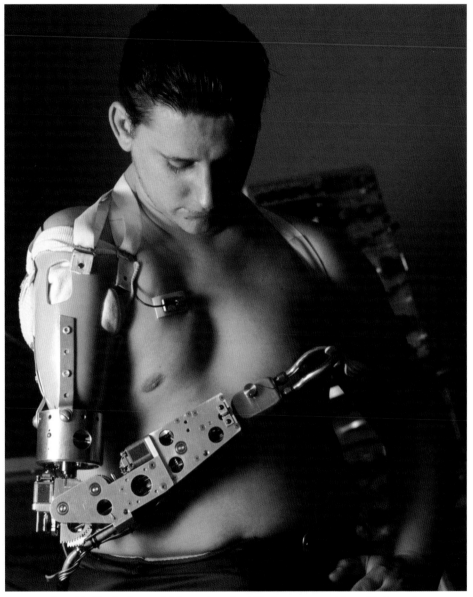

Prosthetic devices, which replace lost limbs, are machines that work in partnership with the rest of the natural body.

These are a few of the ambitious dreams of bionics that might very well come true one day. It is possible that much of what is now thought to exist solely within the realm of science fiction or fantasy will be part of our everyday lives.

CHAPTER 1

HUMAN + MACHINE = BIONICS

Bionics is the science of fusing artificial devices with human body parts to duplicate or improve the body's function. The two main branches of bionics involve artificial limbs and artificial implants. Bionic equipment can enable people with damaged or missing limbs to function at almost full capacity. Bionics can also save lives by replacing diseased or damaged organs with artificial ones. It can restore some hearing to the deaf and some sight to the blind. Beyond all that, bionics can improve human capabilities in otherwise smooth-functioning organs.

Shaping an artificial leg

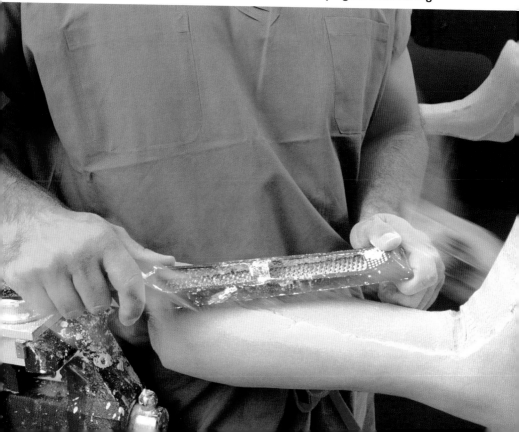

BIONICS BEGINS WITH ARTIFICIAL LIMBS

Bionics has been around for centuries. The Rig-Veda, a sacred poem written in India between 3500 and 1800 B.C., contains the first written account of a replacement body part. The warrior Queen Vishpla loses her leg but returns to battle after being fitted with an iron leg. Today, this kind of artificial limb would be called a "prosthesis."

The ancient Romans of the first century used artificial legs made of wood, bronze, and leather, and artificial arms made of iron. During the late Middle Ages, between the 11th and 14th centuries, knights who lost limbs in battle had prostheses made by their weapons-makers so that they could hold shields and ride in stirrups. Away from battle, those same knights used peg legs and hand hooks that allowed them to perform everyday tasks.

The Renaissance, which began in the 14th century, was a time of growing interest in science, and it brought improvements to the design of artificial limbs. The German knight Götz von Berlichingen (1480–1562) had a mechanical hand made, with joints that could move independently with the help of his good hand. In the centuries that followed, watchmakers and woodworkers used springs, gears, and pulleys to make artificial limbs work better.

In the 19th century, designers of prostheses began to make artificial arms out of wood instead of metal, which made them lighter. In 1812, an arm prosthesis was developed that allowed its wearer to control the movement of the artificial arm and hand. An artificial leg, developed in 1858 by the British doctor Douglas Bly, attached to the remaining part of the user's original limb by suction, and was equipped with joints that enabled an amputee to walk on a level surface.

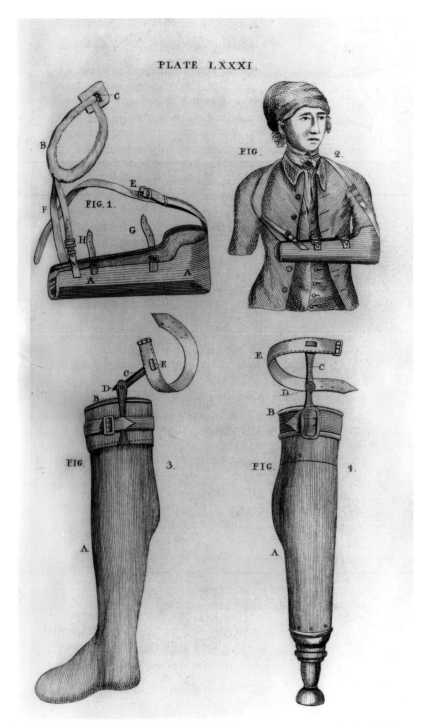

This engraving shows wooden limbs that were common in the 1700s.

"BETTER, STRONGER, FASTER"

The idea of bionics has long been a popular one for writers of science fiction. The current use of the word bionics was introduced by a 1970s television show. Before that, bionics was an engineering term that referred to the study of living things to guide the design of machinery.

The television series, which was called *The Six Million Dollar Man*, centered around the adventures of an astronaut who had suffered crippling injuries in a plane crash. Using experimental techniques, a team of scientists gave the astronaut artificial legs that enabled him to run at sixty miles per hour, an artificial arm of superhuman strength, and an artificial eye equipped with telescopic vision. "We can rebuild him," an official-sounding voice announced in the opening sequence each week. "We have the technology. We have the capability to make the world's first bionic man...better than he was before. Better... stronger...faster."

Television critics did not particularly care for *The Six Million Dollar Man*, and even young viewers noticed that actor Lee Majors had a habit of forgetting which arm was supposed to be

The Six Million Dollar Man

bionic. He often lifted a boulder or a ten-ton truck with his nonbionic arm. Nevertheless, the show was a tremendous hit. It ran for five seasons and spawned action figures, lunch boxes, movie specials, and a spin-off, *The Bionic Woman*. Similar to *The Six Million Dollar Man* was the 1987 movie *Robocop*, in which a police officer, nearly killed in the line of duty, returns as a half-mechanical superman.

Robocop

As computers have become important in modern-day life, other stories about bionics have focused on links between humans and computers. William Gibson's 1984 novel, *Neuromancer*, depicts a future in which people's brains are hooked up to computers, and who undergo major surgery almost as casually as they change their clothes. Gibson coined the word "cyberspace," which has become a popular term to describe the World Wide Web. In the 1990s, human/computer combinations were portrayed, sometimes frighteningly, in movies such as *Johnny Mnemonic* (1995) and *The Matrix* (1999).

MEDICAL IMPLANTS

Although artificial limbs have a long history, the other branch of bionics—medical implants—did not develop until the end of the 20th century. Medical implants are artificial devices inserted in the body to replace specific organs, such as the heart.

One reason it took so long for scientists to develop medical implants is that surgery is necessary to insert the devices, and for centuries, a lack of knowledge prevented doctors from performing operations safely. In fact, surgery was so dangerous that it was done as little as possible. Until the end of the 18th century, surgery was limited to the lancing (puncturing) of boils, the sewing up of wounds, and the amputation of limbs. It was considered such a low art that for a long time it was performed more often by barbers than by physicians.

Medical implants could not be done effectively before the 1900s, when sterile procedures were developed for surgery.

Surgery was extremely risky for several reasons. The chance of infection was high. Not until the mid-19th century did scientists learn that germs could cause infection or keep wounds from healing. Before those discoveries, surgeons often did not even bother to wash their hands between operations. There were no antiseptics and no infection-fighting drugs. In addition, there were no

Louis Pasteur

effective painkillers, so the only way to limit a patient's suffering was for the surgeon simply to work as quickly as possible.

The 19th century brought advances in medicine that made surgery safer and less painful, thus paving the way for medical implants. In 1800, the British chemist Sir Humphrey Davy (1778–1829) discovered that breathing nitrous oxide gas made people insensitive to pain, and suggested that it be used in surgery. In the 1860s and 1870s, Louis Pasteur (1822–1895) in France and Robert Koch (1843–1910) in Germany, working separately, discovered that germs could cause disease. Their discoveries led the

Joseph Lister

British surgeon Joseph Lister (1827–1912) to promote the use of sterilization and antiseptics in hospitals.

As a result of these discoveries, surgery became more common. Then, in the mid-20th century, a major advance in medicine

The heart and lung machine allowed surgeons to operate on those organs while the machine performed the organs' functions for the body.

revolutionized the ability of doctors to perform more complicated surgeries. The heart and lung machine, invented in 1937, allowed surgeons to operate on those organs while the machine performed the organs' functions. Because the machine circulated blood through the patient's body, doctors could perform longer and more difficult procedures.

The miniaturization of motors and complex electronic components also helped lead to workable medical implants. Innovations included the 1948 invention of transistors (devices that open or close an electrical circuit) and the 1958 invention of small electronic devices called integrated circuits.

These two advances enabled engineers to dispense with the bulky collections of vacuum tubes used for electronic switches and create devices tiny enough to fit inside the body. Transistors and integrated circuits eventually gave rise to microprocessors, the still smaller and more powerful computers now found in machines such as personal computers and mobile phones.

Breakthroughs in the field of medical technology in the 1960s further advanced the development of medical implants. The medical laser was a cutting tool that enabled surgeons to work on a microscopic scale. The surgical microscope, with its double

Microprocessors made it possible to create small, powerful, and highly effective machines for use in bionics.

Medical lasers provide doctors with the ability to see very small structures during surgery, which makes it possible for them to insert or work on tiny bionic devices.

eyepiece, let the surgeon see very small structures within the body part being operated on with the medical laser. With these sophisticated new instruments, surgeons were able to work with precision on nerve cells too small to be seen with the naked eye.

Some of the artificial devices that doctors implanted into people's bodies were relatively simple constructions, with few working parts. Hip replacements and shoulder joints are two examples of the simplest kinds of medical implants. Others, such as artificial hearts, were complex machines.

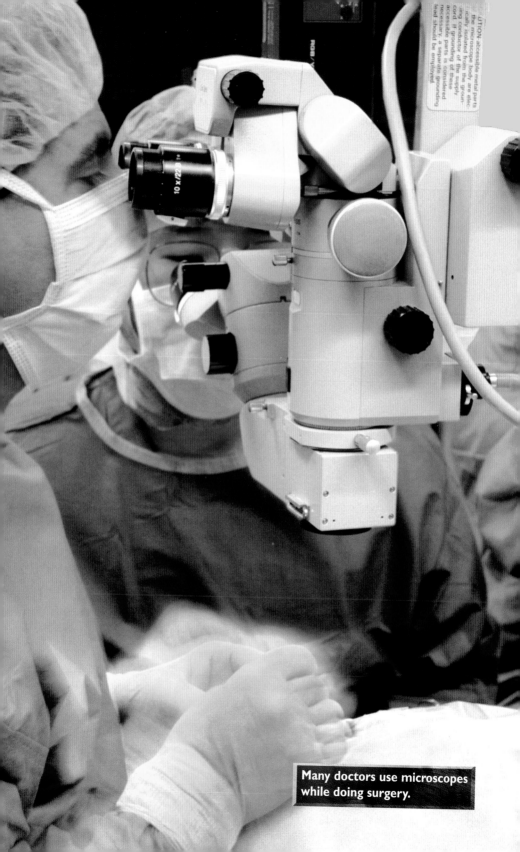

Many doctors use microscopes while doing surgery.

THE CARDIAC PACEMAKER

One of the most successful advances in the field of medical implants is the cardiac pacemaker. First used as an implant in 1958, it has saved millions from premature death or crippling weakness.

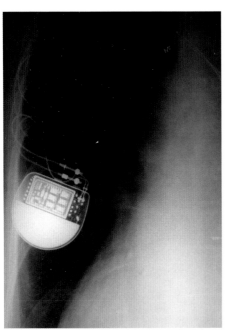

This X ray shows an implanted pacemaker attached to a human heart.

The cardiac pacemaker addresses a special kind of heart condition. The human heart beats around seventy times a minute. Normally it is prompted to beat by the sinoatrial node, a lump of muscle at the back of its right side. Damage to the sinoatrial node can make the heart pump irregularly or too slowly. The creation of the cardiac pacemaker solved this problem by substituting small, painless electrical shocks for the action of the heart's natural pacemaker.

Installing a cardiac pacemaker is a relatively easy procedure. Only a local anesthetic is needed, so patients can stay awake during the operation. A battery is implanted under the patient's skin. A plastic-coated electric wire passes through the veins to the heart, where the wire's bare metal tip makes contact with the heart muscle. Most pacemakers only go to work when the heart itself fails to beat on schedule.

In the years following the introduction of the cardiac pacemaker, medical engineers developed similar electrical devices that helped other organs function. For example, some electrical implants that use the same principles as the cardiac pacemaker take over the job of prompting the lungs to breathe.

ARTIFICIAL HEARTS

Success with devices such as cardiac pacemakers made medical engineers set their sights even higher. Their goal soon became to create a machine that would replace worn-out hearts and thereby save patients from certain death. By the 1970s, surgeons were already performing heart transplants, a procedure in which a person's sick heart is replaced with a healthy heart

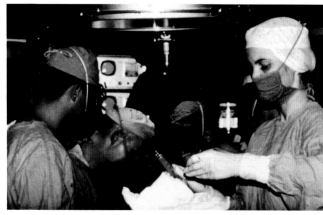

Doctors in Palo Alto, California, performed the first heart transplant.

taken from the body of a recently deceased person. Often, however, these transplants were not successful.

One of the major causes of failure was that the body rejected the transplanted organ. The immune system's job is to attack anything it identifies as foreign, whether the invader is disease-causing bacteria or a lifesaving transplanted organ. The body's immune system would treat the transplanted heart as a threat, and attack it.

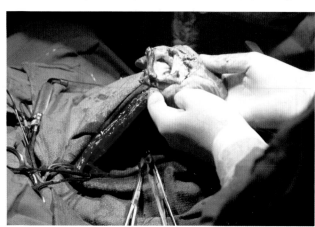

A donor's heart is placed inside the chest cavity during a heart transplant.

THE JARVIK-7 ARTIFICIAL HEART

The Jarvik-7 artificial heart stands out as one of the most impressive achievements in the history of medical implants. It is also one of its biggest disappointments.

The initial work on the first artificial heart dates back to the

Dr. William DeVries

mid-1950s, when scientists began testing it on animals. In 1969, a team led by Denton Cooley of the Texas Heart Institute successfully kept a human patient alive for more than sixty hours with an artificial heart.

The Jarvik-7 artificial heart was developed in the late 1970s by a team of researchers headed by Dr. Robert K. Jarvik. In 1982, a team led by Dr. William DeVries of the University of Utah implanted the Jarvik-7 artificial heart into the chest of a retired dentist named Barney Clark. Clark's heart was failing, and for various medical reasons he could not receive a living heart transplant. The Jarvik-7 was his last hope. Clark survived with the Jarvik-7 heart for 112 days.

Between 1982 and 1985, the Jarvik-7 artificial heart was implanted in five other patients who were facing death from heart failure. All the patients lived longer on the Jarvik-7 than many doctors had expected. The longest survivor was William Schroeder, who survived for eighteen months with the artificial heart.

The results, though impressive, were accompanied by severe side effects. Clark and the others who received artificial hearts developed serious infections and blood clots that damaged their other organs. They died from causes other than heart failure. Ironically, they each died with a Jarvik-7 pumping faultlessly inside them. Although the recipients of the Jarvik-7 lived longer than they would have without their artificial hearts, they had to suffer greatly as a consequence.

In the years that followed, the medical profession became disenchanted with artificial hearts. The technology was simply not yet good enough. Scientists began to use artificial hearts only as a stop gap measure for very sick patients waiting for heart transplants. As of 2002, the Jarvik-7, renamed the CardioWest total artificial heart, is still being used in some hospitals as a "bridge to transplantation."

As the shortage of real hearts for transplants continues and improvements in medical engineering are made, heart specialists often return to the idea of the artificial heart. "In the near future I think that the heart will be the only organ that will be replaced as a totally artificial or manmade substitute for an internal organ," says O.H. Frazier, Chief of Cardiopulmonary Transplantation at the Texas Heart Institute in Houston.

William Schroeder survived for eighteen months with an artificial heart.

CHAPTER 2

BIONICS TODAY

Today, the use of artificial devices to replace injured, worn-out, or missing body parts is standard medical practice. One out of every ten Americans depends on some kind of artificial physical enhancement (not including such commonplace devices as false teeth, eyeglasses, or contact lenses).

Each year, 250,000 Americans receive an artificial knee with the flexibility of a real one. Another 200,000 get a hip joint made of titanium or polyethylene. One hundred fifty thousand people receive cardiac pacemakers. Millions who would otherwise be deaf are able to hear thanks to hearing aids. A new device called the cochlear implant is giving hearing to people who have been deaf since birth. People whose legs have been amputated below the knee

Artificial knee and hip replacements

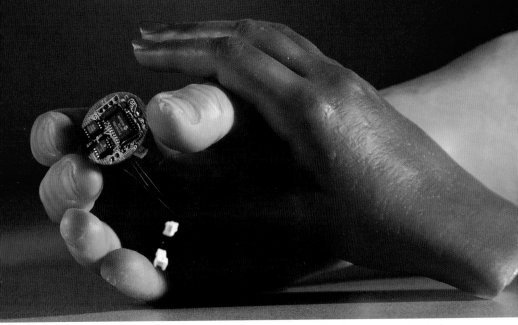

Myoelectric limbs look like natural hands or legs and can be controlled by the actual muscles of the person wearing them.

can run, jump, and participate in physically demanding sports such as skiing and basketball, thanks to advances in artificial limbs.

Many of these developments in bionics have been made possible by breakthroughs in areas other than medicine, especially microprocessor technology (the miniaturization of electronic parts) and electronics. By keeping abreast of these technologies, medical engineers have discovered useful tools that have furthered the science of bionics.

MYOELECTRIC LIMBS

The miniaturization of electronic parts has helped scientists develop artificial limbs capable of making movements directed by the amputee's muscles. The newest state-of-the-art prosthesis is called the myoelectric limb.

A myoelectric limb looks like a natural hand, arm, or leg. Inside are tiny electrical parts, motors, and a place for rechargeable batteries. A myoelectric limb is controlled by the muscles left in the stump of the missing limb. Delicate sensors in the limb pick up electrical signals that are issued from the user's muscles as they contract. Circuitry tells the motor in an artificial hand, for example, to open or close the hand or rotate the wrist, and it even tells it how fast to perform these actions. Myoelectric limbs do not operate with the subtlety of real limbs, however. In addition, users must undergo extensive training to get a myoelectric limb to obey the commands of their muscles.

Another limitation of myoelectric limbs is that they lack a sense of touch. The link between the senses and the brain is one of the subtlest aspects of human biology, and it is the one that scientists have had the most difficulty duplicating.

THE BIONIC EAR

The difficulty in duplicating human senses has made the invention of a device called the cochlear ear implant very exciting to scientists. This implant provides some hearing to people who have been deaf since birth. It is the first bionic device to restore a sense.

Cochlear implants such as this one make it possible to bring hearing to people who were born deaf.

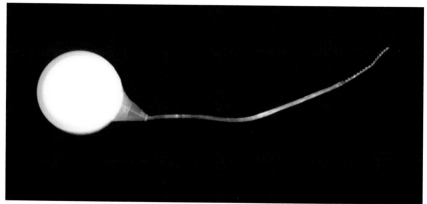

Medical engineers have been working on the cochlear ear implant since the 1950s, but it was not until 1998 that the U.S. government approved it for use in the general population.

The cochlea is a spiral-shaped structure in the inner ear that converts sound vibrations to nerve signals and sends them to the brain. Damage to this organ is a common cause of partial or complete deafness. The cochlear ear implant uses electrodes—tiny wires that conduct electricity—to stimulate the auditory nerves in the same way that the patient's cochlea would if it were functioning properly. Like the cochlea, the cochlear ear implant sends complex signals to the auditory center of the brain that the brain can convert into auditory information. The cochlear ear implant does not restore full hearing, but it enables a person to hear many sounds.

One of the greatest challenges that cochlear implants present to people who have been deaf all their lives is learning to understand what they are hearing. Comprehending speech is a complex, gradual process that begins at birth. Learning to decode language is particularly difficult for adults. Words strung together into sentences will sound like meaningless noise to a person who has never developed the ability to understand spoken language.

THE BIONIC EYE

Giving sight to the blind has turned out to be a more difficult challenge than giving hearing to the deaf. Damage to the retina is a common cause of blindness. The retina is the inner lining of the eye that transmits images through the optic nerve to the brain. Sight is not possible without a functioning retina. Medical engineers and surgeons have begun to treat this problem by implanting an ultrathin microchip into the retina. The microchip does the work of

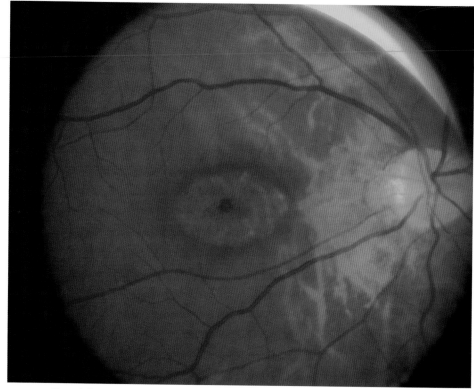
Repairing or replacing a damaged retina, shown above, has been a very difficult challenge to bionics engineers.

the retina and translates the light that falls on the eye into electrical impulses that are passed on to the vision center of the brain.

The amount of vision restored by this new technology is very small. The implant allows patients to see only a bright light shone directly into the eye. For someone who was totally blind before, though, this small step toward the development of true artificial eyes is an important one.

Engineers have also developed an artificial vision system that bypasses the patient's retina altogether. This system uses a TV camera mounted on eyeglasses. The camera sends signals to a portable computer worn in a purse. The computer, in turn, stimulates a one-inch square electrode grid implanted on top of the brain's visual center. The visual center picks up these signals and "sees" what the camera sees.

FUNCTIONAL ELECTRONIC STIMULATION

Bionics today has progressed even beyond the ability to improve hearing and sight. Functional Electronic Stimulation—FES for short—is a technology designed to do the work of one of the most complicated systems of the human body, the nervous system.

The nervous system is a vast network of specialized cells—called nerve cells or neurons—that carries messages from the brain to the muscles and signals to them when and how to move. Since 1780, when the Italian physician and physicist Luigi Galvani (1737–1798) discovered that an electrical current would make an animal's muscles move, scientists have known that the nervous system sends its messages to the muscles by means of electricity.

In a functioning nervous system, the brain directs the muscles with signals that travel through the spinal cord—a cable of nerve tissue that runs from the brain down through the backbone—before they branch out to other parts of the body. When an injury cuts the spinal cord, every part of the body below the site of the injury is paralyzed. The patient cannot feel anything or direct any motion below the point of the injury.

Luigi Galvani

Functional Electronic Stimulation can help restore some movement to those paralyzed by nerve damage such as spinal cord injuries. With FES, the patient or a physical therapist uses a machine to stimulate the patient's muscles with electricity.

To stimulate leg movement, for example, the patient is seated on a stationary bicycle, with his or her feet secured to the pedals.

Electrodes are fastened to the surface of the skin at points that correspond to the muscle. Programmed by a computer to send low levels of current, the electrodes trigger nerve impulses that make muscles contract. In 1983, FES enabled Nan Davis, paraplegic from a spinal cord injury, to rise out of her wheelchair and collect her diploma. For a paraplegic to walk again, even for a minute, seemed like a miracle.

Since then, research on FES has continued, and its usefulness has expanded. FES has had some success in helping disabled people to use their hands, and others to use their bladders normally. Still others are able to counteract one of the side effects of paralysis—the wasting

Parapalegic Nan Davis walks with the aid of a Functional Electronic Stimulation system, which is activated by muscles.

away of muscle—by being able to maintain muscle tone through movement. It is also possible that FES can assist in the regrowth of nerves. Actor Christopher Reeve, for example, who was paralyzed after a fall from a horse, has regained some voluntary movement in his lower body that has been attributed to FES.

For all its advances, FES still has limitations. People with intact nervous systems do not work the muscles at full power because the signals sent to the muscles are subtle. FES makes muscles work at maximum effort, so they tire quickly.

Another limitation is that unlike people with functioning nervous systems, who control their movements through the power of thought, people who use FES control their movements by operating a machine. The ultimate goal of specialists in FES technology is to find a way to control movements with the brain alone. This would involve the implanting of electrodes into those parts of the brain that control the part of the body that is paralyzed—and would, in effect, be the creation of an artificial nervous system. When this technology is successful, paralysis from spinal cord injuries could be a thing of the past.

TODAY'S ARTIFICIAL HEARTS

While some medical engineers work on devices that compensate for a damaged nervous system, others continue to pursue the dream of a replacement for damaged hearts. In 2001, a new kind of artificial heart was approved for human trials (experimental use by selected volunteers). With the right improvements, this artificial heart may eventually be better than a transplant for people whose own hearts have failed.

This latest artificial heart is called the AbioCor. It contains a computer that adjusts the patient's heartbeats. The AbioCor is a major improvement over the Jarvik-7 because it allows its user much more mobility. Most of the Jarvik-7, introduced in the 1970s, was housed outside the patient and connected by tubes. In contrast, the AbioCor is implanted in the same place in the body where the heart is normally located, and the only part outside the patient's body is a battery pack worn on a belt. The battery transmits power to the artificial heart through a waterproof coil attached by an undershirt to the user's chest.

The AbioCor heart is still an experimental device, and therefore by law it can only be implanted in patients who are in imminent danger of death from heart failure—patients whose own hearts are

THE SCIENCE OF THE SMALL

In 1959, the physicist Richard P. Feynman (1918–1988) delivered a lecture to the American Physical Society in

Richard P. Feynman

which he announced his intention to found a new science. This science would be devoted to making things on a microscopic scale — things so small that they could only be seen by a microscope. According to Feynman, it would soon be technically possible to build machines of microscopic size. Using these machines to manufacture ever smaller machines, engineers would one day create machines small enough to grab individual atoms and assemble them like building blocks. Feynman pointed out that this technology could yield many useful applications. Among them were what he called "tiny computers" (Computers were so large at the time of Feynman's lecture that they filled an entire room!), and tiny machines that would be able to enter the human body and repair it from the inside. Besides, Feynman added, the technology would be worth doing "just for fun."

Feynman's lecture has become the founding document of a new science of the small, called "nanotechnology." The prefix "nano" means that something is one billionth the size of something else. According to Ralph Merkle of the Palo Alto Research Center of the Xerox Corporation, current manufacturing methods are "like trying to make things out of Legos with boxing gloves on.... In the future, nanotechnology will let us take off the boxing gloves."

Nanotechnologists point out that living things already consist of microscopic machines. Tiny molecules, too small to be seen with a conventional microscope, do very complicated work in the human body. They store and transport energy, make copies of themselves, and produce proteins. In effect, they are tiny machines. Nano-technologists believe it is possible to build artificial machines that are even smaller and more efficient than these natural ones. Eventually, they feel, it will be possible to manufacture microscopic robots—"nanobots"—equipped with computers and wireless communication, and able to reproduce themselves.

Nanotechnology could have far-reaching effects on bionics. Machines created by nanotechnology would be able to operate on the same microscopic scale as our human biochemistry. They would make it easy to perform the microscopic surgeries needed to link the nervous system with artificial limbs and implants. According to the computer scientist Ray Kurzweil, "nanobodies" could be created that could do work well beyond the capabilities of our biological immune systems. Kurzweil says that these nanobodies could be "launched into our bloodstreams" to "seek out and destroy pathogens, cancer cells...and other disease agents." He predicts that one day, "we will be able to reconstruct any or all of our bodily organs and systems."

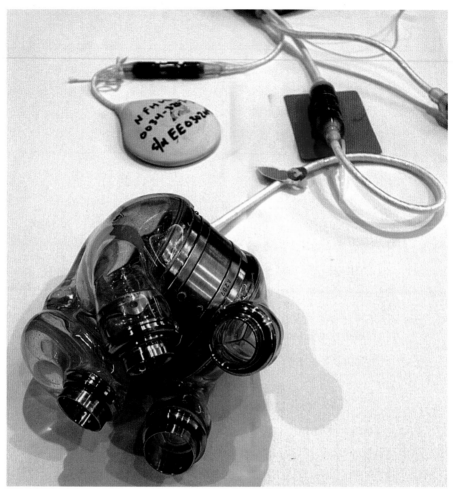

The AbioCor is the latest advancement in artificial heart technology. It contains a computer that adjusts the patients heartbeats and is mostly contained inside the body.

not expected to function for more than thirty days. The stated goal of the first human trial of the AbioCor artificial heart was to extend the users' lives by sixty or more days. The trial has already met this goal. To date, the longest a patient with the AbioCor has survived is 151 days. This is still a long way from the eighteen-month record of the Jarvik-7, but scientists hope that additional advances will eventually improve the performance of the AbioCor. With every new development, the science of bionics comes closer to breakthroughs that are not quite within reach today.

CHAPTER 3

BIONIC FRONTIERS

Much of bionic research today is being conducted to create technology that will enable people to control artificial limbs as naturally as if the limbs were their own. Scientists hope that in the future, people will be able to feel with artificial fingers, and see with camera eyes. The technology to make this happen is not yet a reality. To make artificial muscles work like real ones, and to make artificial vision as acute as natural vision, scientists must find a way to get nerve cells to interact as well with machines as they do with the body's natural biological equipment. Once this engineering challenge is met, many bionic dreams will be within reach.

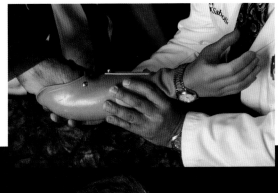

Today's bionic limbs are far more advanced and sophisticated than those of only a few decades ago.

NERVE CELLS AND ELECTRODES

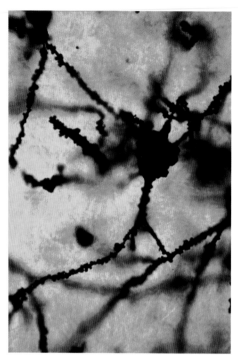

Close up of a nerve cell

The nervous system enables muscles to function and people to see, hear, taste, touch, or smell. Nerves make human function possible by carrying precise, detailed messages to and from different parts of the body.

There are two obstacles to overcome in this area of bionic technology. First, surgeons must be able to perform the intricate microsurgery needed to hook up tiny, complicated nerves to machine sensors. Second, medical researchers must find ways to prevent implanted electrodes and nerve cells from damaging each other.

Electrodes are made of metal wire. Metal is hard; nerves are soft. The hard metal can easily damage the soft nerves. When surrounded by flesh and blood, metal electrodes tend to corrode and release toxic chemicals into the body.

Solutions to these problems are already in the works. Some scientists are attempting to make electrodes out of materials that will not damage nerves. They are also working to develop so-called biomaterials that will chemically interact with the body's cells and tissues. Physicists at the Max Planck Institute of Biochemistry in Munich, Germany, have made a microchip of silicon that can directly stimulate a nerve cell. (Microchips, also called "chips," are computer devices that are made on a very small scale.) The physicists chose silicon, a nonmetalic material, because it is not prone to the same kind of corrosion and toxicity as metal.

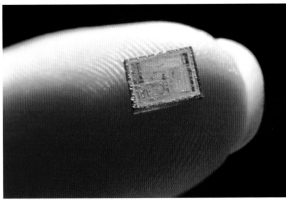

Computer chips like this one make it possible for today's bionic devices to perform highly complex functions.

Scientists are also looking at ways that nerve cells and machines might communicate without touching at all. One machine that has long been used in hospitals, the EEG (electroencephalograph), might make this possible one day. An EEG consists of a group of wires attached painlessly, with a gel, to a patient's skull. The wires are connected to a computer and pick up electrical fields that are produced whenever nerve cells carry a signal. Doctors use EEGs to diagnose brain problems such as sleep disorders and epilepsy (a disorder that results in seizures). An EEG does not touch nerve cells, so it cannot damage them.

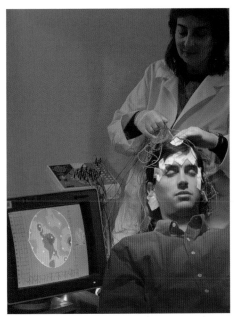

EEGs are playing an important role in helping scientists figure out how nerve cells and machines may better communicate in the future.

Since EEGs can read brain waves without touching nerve cells, scientists believe that they will one day have a useful application for bionics. EEGs today are capable of sensing only large-scale brain activity. If these machines could be made sensitive enough to pick up all brain signals, then bionic arms and legs—and any machine, from a garage door opener to an airplane—could be directed by the power of thought alone. EEGs could enable people to move their bionic parts just as easily as they are able to move their own natural biological parts.

BRAIN IMPLANTS TODAY

Despite the challenges of joining nerve cells and electrodes, scientists have begun to experiment with devices surgically implanted in the brain. In 1998, neuroscientists at the University of Teubingen in Germany and Emory University in Atlanta, Georgia, tested a device implanted in the brains of paralyzed stroke victims. The implant picked up changes in brain waves and transmitted the changes to a computer. After intensive training, the patients were able to use their brain waves to move a cursor on

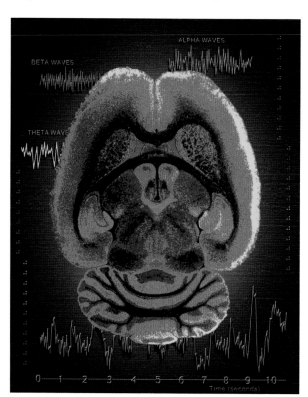

This colored EEG print out shows brain wave activity in a healthy human.

the computer screen. The patients used the cursor to spell out words on the screen.

In 2001, researchers at Duke University conducted a bold experiment that shows the capabilities that bionic brain implants may one day allow. The researchers implanted a bionic device consisting of electrodes into the brain of a monkey, and then linked the bionic implant wirelessly up to the Internet. Next, they directed the monkey to perform tasks that

Bionic implants currently exist that can read the brain waves of paralyzed people and transmit those waves to a computer in order to spell words or perform other functions.

involved reaching, an action that caused the monkey's brain to emit signals that were picked up by the electrodes implanted in its brain. These electrodes, hooked up to the Internet, sent signals to the arm of a robot located in a research facility 600 miles away. The signals were able to move the robot's arm.

Despite the success of these experiments, communication between nerves and machines is still many years away. The ability to control a cursor with one's brain is a long way from the ability to see with artificial eyes. The ability to control a robot arm is a lot simpler than the ability to control a limb and have it feel like one's own. Once the problem of getting nerve cells and machines to communicate with each other is solved, though, superhuman accomplishments could follow quickly. One day, it may be possible to communicate across great distances—as people do today by dialing a phone or tuning a radio—simply by thinking. Today, people can see through walls with the help of infrared cameras.

Someday, equipped with an implant, people might look through walls simply by squinting.

FUTURE POSSIBILITIES

The promise of bionics, though, goes beyond the mere enhancement of present-day capabilities. In the future, people may actually have more than one body. One would be a flesh-and-blood body, with a brain implant equipped to communicate wirelessly with computers. With the implant, people could direct the movements of computer-controlled robot-bodies that would be activated as needed. People could sit at home and operate their robot-bodies from a distance. They could see what their distant robot-bodies see, hear what robot-bodies hear, and manipulate these external bodies as easily as people now talk on the telephone. They might use their robot-bodies to fight wars, explore outer space, or visit friends thousands of miles away in a single day.

Robotic arms, such as these, can be operated from remote locations and are capable of performing increasingly complex and delicate functions.

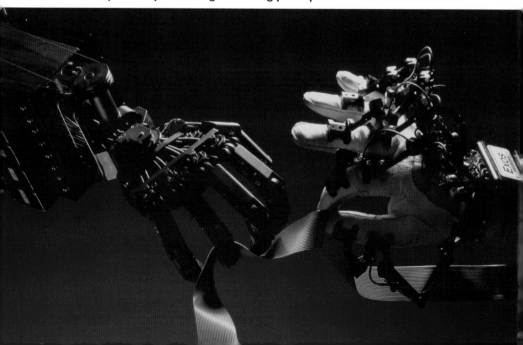

These extra bodies might be machines with a human shape, or they might look quite unlike human bodies. They could have human senses, though, and they could possess other attributes, such as infrared vision, radar, or sonar.

If computers could receive information directly from human brains, a person might be able to download his or her entire mind into a machine—thereby making a copy of a person much as one

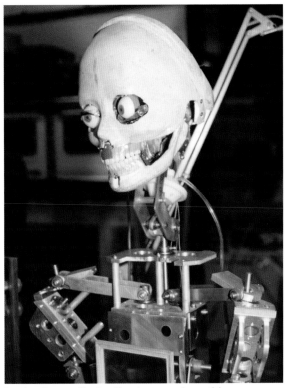

In the future, "cybot" robots such as this one may be operated remotely or through the transmission of brain waves.

might make a copy of a computer file. The possibility raises some interesting questions. Would the copy in the computer be the same person as the original? What would its rights be? Would it have feelings and desires? Would it claim to be the original? If the person dies, does he or she live on in the computer? Where does the human end and the machine begin?

For many, these questions raise troubling issues. If it ever becomes possible to copy a person's mind, many people faced with death will be tempted to do so. It would be the only chance they would have of continuing to exist. Whether or not there could ever be a bionic mind, people are sure to speculate about the possibilities.

EXOSKELETONS

Global positioning satellite

What if it were possible to put on a mechanical suit that would vastly increase strength and agility, that would make a person invulnerable to bullets and radiation, and perhaps enable him or her to fly? For many, this would be the fulfillment of a fantasy. The U.S. government is serious enough about the possibility to have invested $50 million in it. In 2000, the U.S. Defense Advanced Research Projects Agency (DARPA) asked private companies to write proposals for "human performance augmentation systems."

The project is called "Exoskeletons for Human Performance Augmentation." In nature, an exoskeleton is the name given to the hard outer structure that supports the body of a creature such as a crab or insect. Instead of having an internal skeleton, this kind of animal has an external skeleton that serves as a kind of armor.

The exoskeleton that DARPA envisions would be a lot more than armor, though. It would be a wearable machine that could turn ordinary soldiers into supertroops. Soldiers equipped with DARPA's exoskeletons could run faster, jump higher, and lift heavier objects than they could without the suits. They could be equipped with jetpacks that would allow them to fly for short distances. The machines would have special features, including the Global Positioning System (GPS), which uses satellite transmission to indicate one's location at any time. Some of the suits, designed for special missions, would offer protection against hazards such as projectiles, radiation, or toxic agents. They may use special computerized fabrics that will monitor a soldier's heart rate and breathing rate. They might even be equipped with artificial intelligence and subject to remote control. If a soldier is seriously wounded, the exoskeleton could be given a command to take over and bring him or her home safely.

Different teams are currently at work on different aspects of the exoskeleton project. What exoskeletons will actually do, and when, depends on the results of these individual lines of research. DARPA estimates it will be another ten years before testing begins on this new technology.

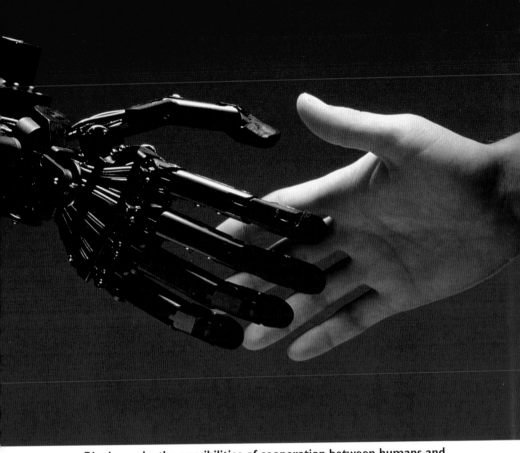

Bionics make the possibilities of cooperation between humans and machines nearly limitless for the future.

A BIONIC WORLD?

Bionic possibilities are exciting because technology has always helped human beings overcome limitations. In a way, tools are extensions of the human body: Using forklifts and cranes, humans are able to lift objects that weigh tons. Airplanes enable people to fly. It is possible, though, to dream of lifting tons with one's bare hands, and to fly without an airplane. The special appeal of bionics is its potential to fulfill such dreams. That is why to so many, the linking of machines to the human nervous system is an irresistible idea. How far will this idea be taken? Only time will tell.

GLOSSARY

AbioCor heart An artificial heart first tried on human patients in 2001.

blood clot A mass of coagulated blood that consists of red blood cells, white blood cells, and platelets entrapped in a fibrin network.

chip A small piece of semiconducting material (usually silicon) on which an integrated circuit is embedded. Sometimes called a "microchip."

electrodes Electrical components, usually wires, that detect electrical activity in nerves.

download To receive a computer file or program via cable or telephone lines.

implant An artificial part designed to be placed inside the human body.

integrated circuits Small electronic devices made of a semiconducting material. Also called "microchips" or "chips."

laser A medical instrument that produces a powerful beam of light and can produce intense heat when focused at close range. Lasers are often used in surgery to vaporize damaged cell tissue.

microchip A unit of packed computer circuitry (usually called an integrated circuit) that is manufactured from a material such as silicon on a very small scale. Sometimes called a "chip."

microprocessor A computer processor on a microchip. It is the "engine" that goes into motion when a computer is turned on. A microprocessor is designed to perform operations that involve arithmetic and logic.

microsurgery Extremely delicate surgery done with the help of a microscope.

miniaturization: The technological advancement that makes it possible to design smaller and smaller machine components; as a result of miniaturization, components take up less space, are faster, and require less energy.

neurons Cells of the human nervous system.

nervous system The system of the body which sends and receives messages by means of electrochemical signals to control thought and movement. The nervous system includes the brain.

physical therapist A specialist trained to use exercise and physical activity to condition muscles and improve the patient's level of fitness.

processor The circuitry that processes basic instructions which drive a computer. The processor in a personal computer is embedded in small devices and is often called a microprocessor.

prosthetics The surgical specialty concerned with artificial devices.

retina The part of the human eye that converts light into nerve signals.

semiconductor A material that is neither a good conductor of electricity (such as copper) nor a good insulator (such as rubber). The most common semiconductor materials are silicon and germanium.

silicon The basic material used to make computer chips, transistors, and other switching devices. Its atomic structure makes the element an ideal semiconductor.

surgical microscope A microscope with double eyepieces for vision with both eyes that is used to see very small structures, like nerve cells, that are being operated on.

transistor A device composed of semiconductor material that amplifies a signal or opens or closes a circuit.

FOR FURTHER INFORMATION

Books
Sarah Angliss, Ian Thompson, Stephen Sweet, *Superhumans: A Beginner's Guide to Bionics*. Brookfield, CT: Millbrook Press, 1998.
This book looks at possible future methods for prolonging life, such as genetic engineering and the use of artificial body parts. Discusses robots, artificial intelligence, cybernetics, cloning, and other technologies.

Thomas H. Metos, *Artificial Humans: Transplants and Bionics*.
New York: J. Messner, 1985.
A general account that covers artificial limbs, organs, bones, skin, eyes, and ears.

Websites
Marshall Brain's How Stuff Works
"How Artificial Hearts Work" by Kevin Bronser
http://howstuffworks.lycoszone.com/artificial-heart.htm

Scientific American Frontiers: Guide to Superhumans and Bionics
http://www.pbs.org/safarchive/4_class/45_pguides/pguide_401/4541_idx.html

CNN.com
Ethics Matters: Who Needs Bionics? Recycled Humans are Here
http://asia.cnn.com/HEALTH/bioethics/9907/body.transplants/

ABOUT THE AUTHOR

Maxine Rosaler is a writer who lives in New York City.

INDEX